W9-CYE-630

the
EXTINCT SPECIES
collection

THE
DODO

For a free color catalog describing Gareth Stevens Publishing's list of high-quality books and multimedia programs, call 1-800-542-2595 (USA) or 1-800-461-9120 (Canada). Gareth Stevens Publishing's Fax: (414) 225-0377.
See our catalog, too, on the World Wide Web: http://gsinc.com

Library of Congress Cataloging-in-Publication Data

Green, Tamara, 1945-
     The dodo / by Tamara Green ; illustrated by Tony Gibbons.
        p. cm. — (The extinct species collection)
     Includes index.
     Summary: Describes the physical characteristics and habits
  of the extinct dodo, a large flightless bird that lived on the
  islands of Mauritius and Réunion in the Indian Ocean.
     ISBN 0-8368-1590-4 (lib. bdg.)
     1. Dodo—Juvenile literature. [1. Birds. 2. Extinct animals.]
  I. Gibbons, Tony, ill. II. Title. III. Series.
  QL696.C67G74   1996
  598.6'5—dc20   96-5002

First published in North America in 1996 by
Gareth Stevens Publishing
1555 North RiverCenter Drive, Suite 201
Milwaukee, Wisconsin 53212 USA

This U.S. edition © 1996 by Gareth Stevens, Inc. Created with original © 1995 by Quartz Editorial Services, 112 Station Road, Edgware HA8 7AQ U.K.

Additional artwork by Clare Heronneau

U.S. editors: Barbara J. Behm, Mary Dykstra

Printed in Mexico

1 2 3 4 5 6 7 8 9 99 98 97 96

the
EXTINCT SPECIES
collection

THE
DODO

Tamara Green
Illustrated by Tony Gibbons

Gareth Stevens Publishing
MILWAUKEE

Contents

Meet the dodo

Most birds can fly. But there are a few that cannot, including the ostrich, the penguin, and the kiwi. Also unable to take to the air was the **dodo**, a bird that is now extinct.

The **dodo** was originally known as the *valghvogel* (<u>VAL</u>-FO-GHEL), which means "disgusting bird." But did the **dodo** really deserve this rude name?

What have scientists been able to discover about the **dodo** from its few remains? Where did it live?

Sailors first sighted live **dodo** birds in 1598. But, just one hundred years later, the **dodo** had become extinct. The **dodo** may be dead, but the fascinating story of its existence lives on.

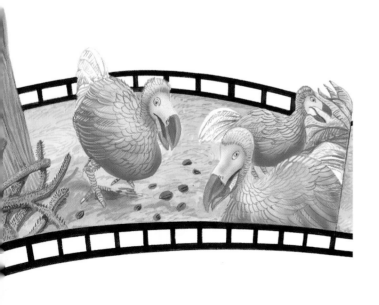

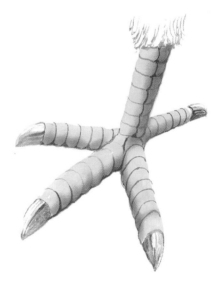

Flightless

No one is quite sure how the **dodo** got its name. One possibility is that it was from a Dutch word *duodu* (DOO-<u>OH</u>-DOO), meaning "idiot," that was used by early sailors.

Perhaps these European travelers thought this bird was not very smart because it could not fly. It could not even run very fast, as its large belly almost scraped the ground. Then again, the **dodo** may have been named after its call. (Try saying the word *dodo*, and see if you can make it sound like a bird call.)

There are several written accounts dating from the seventeenth century that describe the **dodo**. Its main feature was a large, hooked beak. This was probably its main form of defense, since it had to stay and face the enemy rather than fly away. The **dodo**'s bite was quite painful to a victim.

bird

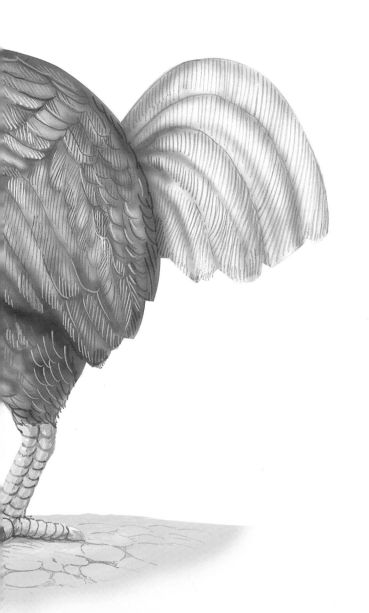

The **dodo**'s body was round and plump. The bird had a thick and sturdy neck. Its tail feathers were only decorative. Its fluffy wings did not have a practical function either.

The **dodo** measured about 39 inches (100 centimeters) from the tip of its beak to the end of its tail. Its eyes were large and protruding and looked like diamonds. The adult birds were generally gray, with white tail feathers, yellowish legs, and green or black beaks.

The **dodo** was once thought to be a form of ostrich or vulture. However, in the mid-nineteenth century, a Danish professor concluded that the **dodo** was actually a large type of pigeon or dove. It had lost the ability to fly mainly because its body became too heavy and its wings too small and weak.

Indian Ocean

The islands of Mauritius and Réunion in the Indian Ocean were home to the **dodo**.

Both islands have a warm climate throughout the year, as was the case when the flightless **dodoes** roamed their shores. The **dodoes** probably nested at ground level on the sides of cliffs.

The remains of **dodoes** have also been found in caves on another Indian Ocean island, Rodrigues. Scientists are intrigued by this discovery.

habitat

The discovery in the caves may indicate that the birds found shelter from the heat of the day or stormy weather. Or perhaps these remains were left by hungry human predators.

The **dodoes** of Mauritius seem to have been more sociable than those of the other two islands. On Mauritius, as shown here, the **dodoes** lived in groups. Elsewhere, the birds lived alone.

Hunted

Early settlers on Mauritius and other Indian Ocean islands are said to have brought about the extinction of the **dodo**. It is believed the settlers hunted the **dodoes** for food. As a result, the **dodo** population quickly declined.

In addition, although some **dodoes** lived near the coast, most inhabited thick forests. These areas were not generally explored by sailors and settlers. Therefore, it is unlikely that humans were the only predators.

But are humans completely to blame? It is probably true that the **dodo** was eaten by people, but it is unlikely the birds were killed in massive numbers for food.

What is far more likely is that predatory animals were brought to Indian Ocean islands by the settlers. These animals were allowed to prey on what they wished.

creature

These free-roaming creatures — probably pigs, dogs, rats, monkeys, and cats — may have hunted and eaten **dodoes**.

The **dodo** may have given a few nasty nips with its beak to any predator.

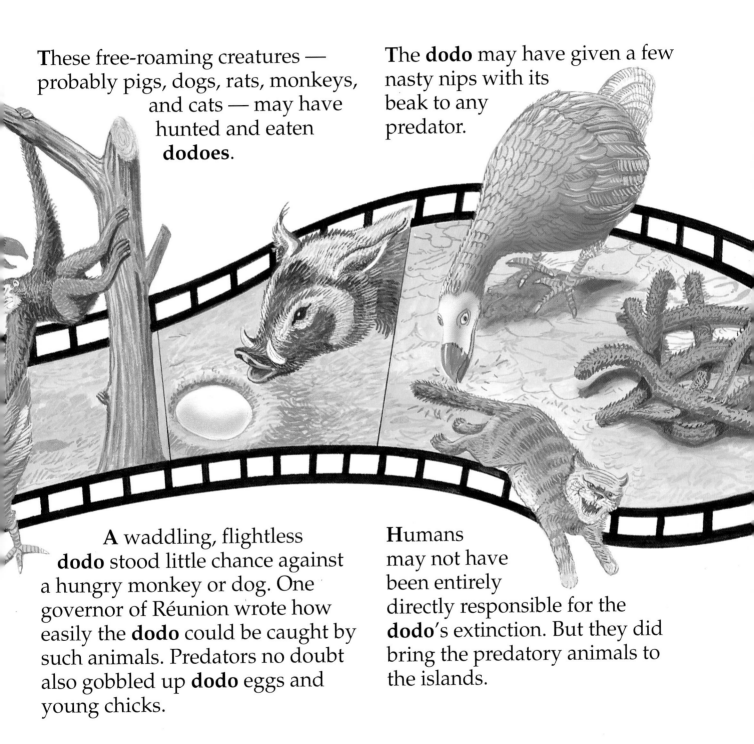

A waddling, flightless **dodo** stood little chance against a hungry monkey or dog. One governor of Réunion wrote how easily the **dodo** could be caught by such animals. Predators no doubt also gobbled up **dodo** eggs and young chicks.

Humans may not have been entirely directly responsible for the **dodo**'s extinction. But they did bring the predatory animals to the islands.

11

The dodoes of Réunion Island

Like Mauritius, Réunion is an island in the Indian Ocean. It, too, was once home to a type of **dodo**.

The Réunion **dodo** was about the size of a turkey. It was very tame and heavy. These **dodoes** lived mostly on their own in secluded places. Their plumage was usually white or sometimes yellow.

Most likely, the **dodoes** of Mauritius and Réunion evolved from the same species, but in slightly different ways.

In 1699, one traveler sent two **dodoes** from Réunion to the king of France. But as soon as the birds were taken onto the ship, they became unhappy. They refused to eat or drink. It was reported that the **dodoes** cried because they were so homesick. The birds got terribly thin and eventually died of sadness and starvation.

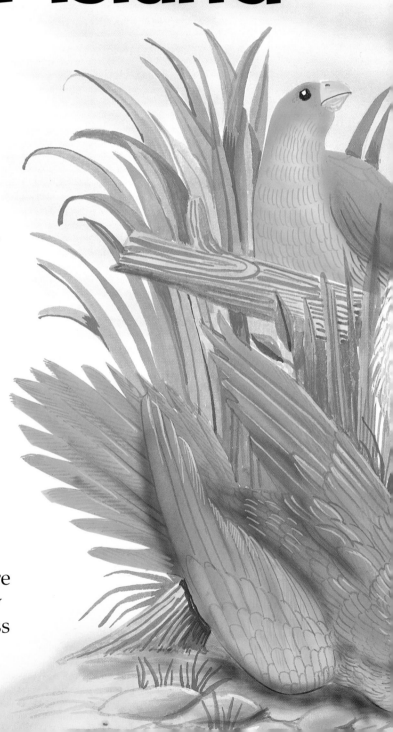

13

A new theory

Recently, a group of British scientists were reconstructing a model of a **dodo** for an exhibition. They started by studying early drawings of the bird that were made on Mauritius during the seventeenth century.

Most of these illustrations, they noticed, show the bird as being thinner than the **dodoes** depicted in European paintings. European artists had painted portraits of the birds from looking at live specimens that were brought to Europe.

The scientists, therefore, concluded that **dodoes** may not have been as plump in their natural habitat as they became in captivity. This picture, for instance, shows a stout **dodo** brought to London over three hundred years ago. **Dodoes** in the wild were probably much thinner and even quite graceful.

about the dodo

Alice meets

The **dodo** was also a fictional character in the book *Alice in Wonderland* by Lewis Carroll. Carroll wrote about his fantasy **dodo** after seeing the remains of a real one when visiting a museum in Oxford, England.

In his black-and-white drawings for *Alice in Wonderland*, artist Sir John Tenniel drew the fantasy **dodo** with a walking stick. He also gave the **dodo** hands. Pictured is a color illustration based on Tenniel's work.

At one point in Carroll's story, a mouse, duck, eaglet, lory (a type of parrot), and several other creatures are all very wet because they have fallen into a pool. The **dodo** suggests they have a *Caucus-race* (a nonsense word) in order to dry off.

the dodo

The **dodo** proceeds to mark out a circular course. It is a very strange one, however, because there are no start or finish lines. There are no rules, either. And the competitors, including Alice, can each start the race whenever they like.

After about thirty minutes, they are all dry, and the **dodo** declares the race over. No one is quite sure who has won — after all, it was hardly an ordinary race! After pausing a moment to think, the **dodo** declares, "*Everybody* has won, and *all* must have prizes."

However, there are no prizes to be seen. Fortunately, though, Alice has some candy in her pocket, and the creatures get one piece each. But there is none left for Alice. The **dodo** suggests she be awarded a thimble that is already hers!

The story of

It's a very odd fact but ever since the **dodo** became extinct on Mauritius, there have been no new trees of a certain species there.

This tree, *Calvaria major* (CAL-<u>VAR</u>-EE-AH <u>MAY</u>-JOR), was commonly known as the **dodo** tree. Scientists think the **dodo** ate its fruit, crushing the seeds during digestion.

This apparently helped the seeds germinate, or sprout, so new **dodo** tree saplings could grow. Without the **dodo**'s help, germination stopped. This unfortunately seemed to mean that when the few remaining **dodo** trees die, the entire tree species will become extinct — just like the **dodo** bird.

But a scientist has come to the tree's rescue. He has fed the **dodo** tree fruit to turkeys that digest the fruit in the same way as **dodoes**. As a result, a few seeds have sprouted. Scientists hope more **dodo** trees can be grown from these sprouts.

the dodo tree

Legends about

Most of the **dodoes** that are exhibited in museums throughout the world are models constructed from plaster and the remains of other birds. Genuine remains of **dodo** skin and feet can be seen in England's Oxford University Museum; in the British Museum in London, England; in the city of Prague in the Czech Republic; and in Copenhagen, Denmark.

Along with these remains, many legends and stories about **dodoes** have also survived. For example, there are stories about how **dodo** meat tasted. One writer described the Réunion **dodo**'s flesh as absolutely delicious. In contrast, a Dutch sea captain reported the meat of the **dodo** as being tough, smelly, and fatty.

Another historian described how the keeper of a **dodo** exported to London gave her bird not only leaves and seeds, but also a good supply of stones to eat. Scientists think **dodoes** swallowed stones to help with digestion.

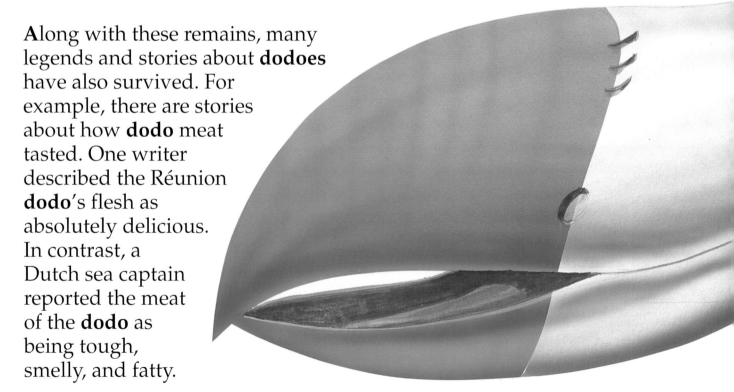

the dodo

Numerous bones of birds resembling the **dodo** have been found in caves on the island of Rodrigues. It is possible that the birds took shelter in the caves or that they were caught and brought to the caves to be eaten — by either human or animal predators.

Perhaps the most bizarre legend handed down about the **dodo** is that, like humans, it would actually shed tears when cruelly treated or when it felt pain.

Dodo data

The **dodo** probably became extinct on the island of Mauritius by about 1690. It was undoubtedly one of the most extraordinary creatures ever to have walked on the planet.

From descriptions dating back to the time when this flightless bird lived, the Mauritius **dodo** had grayish plumage. The Réunion **dodo**, however, was far paler.

When fully grown, the **dodo** probably weighed about 50 pounds (23 kilograms). It did not need to be light in weight, since it could not fly. Its beak could be used for defense. The **dodo**'s feet had four clawed toes.

Dodoes ate seeds and leaves they picked off the ground.

One early writer reported that the dodo had a large gizzard (part of a bird's stomach where hard food is broken up). In the gizzard, he found a flattened stone about the size of a chicken's egg. Dodoes swallowed stones to help them digest their food, much like chickens and pigeons. Some dinosaurs did the same thing — Brachiosaurus (BRACH-EE-OH-SAWR-US), for example, which lived long before the dodo.

For the most part, dodoes lived on their own, as solitary creatures, until mating season. Like some other species of birds, dodoes mated for life. They nested on the ground because they could not fly. The female usually laid just one white egg at a time. Dodoes were particularly good at caring for their young. If the female saw another bird approaching the young, she would call for her mate to shoo it away. Likewise, if the male saw another bird coming near the young, he would call to the female to help drive it off.

Glossary

Brachiosaurus — a huge dinosaur with a bulky body and a long, thin neck that lived many millions of years before the **dodo**. Its remains have been found in Africa and North America.

dodo tree — a tree on the island of Mauritius with the scientific name of *Calvaria major*. Its existence has been threatened since the extinction of the **dodo**.

germination — the process by which a seed begins to grow into a plant.

gizzard — the area of a bird's stomach where food is broken up.

Mauritius — a large island in the Indian Ocean where the **dodo** lived peacefully until the arrival of settlers.

Réunion — an island in the Indian Ocean that was once home to a particular type of **dodo**.

Rodrigues — a small, isolated island 300 miles (480 kilometers) away from Mauritius. Rodrigues, Réunion, and Mauritius form the Mascarene Islands.

valghvogel — the original name for the **dodo**, meaning "disgusting bird."

Index